MOON BOOK
Emily Eve Weinstein

MOON BOOK
Emily Eve Weinstein

A pictorial journal of a year
of painting the full moon

Point Pleasant, West Virginia

Library of Congress Catalog Card Number: 99-62140
ISBN: 0-9667246-1-5

First Edition, First Printing

Printed in Hong Kong

Book and Cover Design: Mark Phillips
Art Photography: Jennifer Collins
Artist Photographs: Alex Harris
Jacket Text: Wendy Hower (adapted from *The News & Observer*, Raleigh, North Carolina)

Discovery**Press**

P.O. Box 304
Point Pleasant, WV 25550
mdg@eurekanet.com

www.moonbook.com

Contents

Introduction

I yearn to capture every full moon. Be in sync with its essence. By doing commissioned art work for twenty years I've gotten out of sync with my own heart as I've tried to please others. With this series I'm taking a journey through my painting, following where it leads. Making the time, abandoning my business, five days a month.

I've discovered the exhilarating freedom of mucking around with oils and have endured an acute breathing disorder caused, I thought at the time, by the fear of freezing while painting outdoors. The images in this book reflect those extremes as they go from tight to loose… direct from life, to nowhere I'd ever been before.

As I sit under the moonlight, sometimes scared of
being mugged, often bitten by mosquitoes or sleep-
deprived, I distance myself from the scene and think,
"Here sits our heroine in these hostile circumstances."
I'm coming to realize that we are all villains and
heroines starring in our own individual life stories.
The planets may have some divine rights over us,
but it is what we do with our lives that makes either
poison or magic.

I hope your reading and viewing this yearlong odyssey
will touch something in your life — and you are
inspired to find some magic of your own.

Acknowledgments

I wish to thank

Bonnie Campbell

Jennifer Collins

Patrick Grace

Alex Harris

Scott Hines

Wendy Hower

Kitty Lynn

Mark Phillips

Nancy Sears

Ron Shores

Frances Vega

My parents & sister

Dedicated to active environmentalists

JUST UP

9/16/97 — 9″ x 6.5″

Starting out this series, I forgot where I placed the greens and reds on my cookie-tin palette. The yellow and white can be seen in the moon's light, so I am painting the contrasting shapes I can distinguish.

LOOKING THROUGH THE WOODS
9/16/97 — 11˝ x 8˝

Mushy and bold, this painting reflects nothing of what I've done in my 20-year career as a freelance artist: meticulous, realistic, time-consuming. This is the liberation I've yearned for. Sitting in the front yard, I duck as a car goes by. I feel strangely exhilarated.

HANGING ROCK

10/15/97 — 8.75˝ x 10.5˝

All week I've been trapped indoors doing a mural. This is a quick escape to Hanging Rock State Park, despite murky weather. At the beginning of the trail a wholesome, athletic-looking guy asked to see the painting I had done in the rain. He was impressed enough to re-hike the trail with me. The sun comes out. Jon sunbathes as I paint the view. A miserable day is turning pleasurable.

The camp's supervisor turns out to a be a painter and very enthusiastically wants to check out this night-painting stuff, so he packs up some gear and joins me. He gets cold and leaves saying he'll try this again at his home in Florida.

MOON OVER ROCK
10/15/97 — 7˝ x 9˝

THE TUCKERS' CAMPSITE

10/15/97 — 11″ x 8″

I really wanted to climb
back up Hanging Rock,
but laziness, the law,
and slippery rocks made
me rethink that idea.
Instead I am painting the
neighboring campsite.
They are retired Hell's
Angels doing the "family
thing" since the arrival
of their blue-eyed angel,
now four.

Truck in Moon Light
10/15/97 — 9˝ x 9˝

The dog is snug and asleep in the truck. I am invigorated by the day's rain, painting, flirtation, new friends, and so I decide to do a moonlit portrait of my '91 Nissan.

BEFORE LEAVING
10/16/97 — 8.5″ x 6″

Home beckons, but before leaving I walk up a short trail to capture one last image. The best vacations are often the shortest.

PLANTS' LAST NIGHT OUT
10/22/97 — 11″ x 9″

They are about to come in for the winter, a little early as my 81-year-old landlord will be power-washing tomorrow. They are all gardenias rooted from the same plant given to me five years ago by a student.

Autumn Lights
11/4/97 — 9″ × 11″

Heading north on Kerley Road. A couple of years ago the fall colors were absolutely spectacular. This year, the trees are showing the loss of so many friends and relatives. Hurricane Fran devastated the land, taking out zillions of tall trees. The loss is still felt by those left standing.

THE CREEK
11/10/97 — 12″ x 9″

It's been kind of tricky
capturing the season as
it's been wet and
miserable. The dog, Daisy
Mae, wants me to finish
up so we can go home.
She hasn't even tested the
water.

BILL POPE'S STUDIO
11/14/97 — 10″ x 7″

Clouds, rain and fog.
The moon finally appears.
The artificial and natural
light do well together.
It's nice when the
manufactured works in
harmony with nature.

ROGER MANLEY'S

11/15/97 — 10" x 8.5"

It's the night after the full moon and I'd hoped to get something close to it being full. Instead it has drizzled on everything: the fire, the guests, Roger's peach cobbler, Kali, his dog. I sit on a bench doing a thick gooey quick painting that somehow manages to capture a corner of "the scene."

DOWNTOWN

11/22/97 — 10″ x 8.5″

Trying to find more occasions to spend with the boyfriend, I suggest he sit with me while I paint a night scene. After two minutes of our searching for a vista, Julian abandons the idea — and me.

THE SNOW BUILDING

12/12/97 — 12″ x 9″

The lawyer who rents this
penthouse knows me only
casually, yet he leaves a
key while he is away.
You know, I never found
a light switch, so to
illuminate what I am
doing I grapple open
the refrigerator door and
view the painting in its
chilly light.

COLD NIGHT
12/14/97 — 12″ x 9.5″

There is a beautiful,
almost-full moon, but I'm
painting inside tonight as
my breathing has become
labored. I think it is worse
when it is cold outside.
So this is the apartment
complex across from
Julian's place, done from
memory and imagination.
Not painting from life
feels like cheating, not
part of the rules, but as I
go along with this project,
conditions form the work
and so it takes on a life
of its own.

ACROSS THE FIELD
12/22/97 — 9″ x 10.5″

One of those intense sunsets that appear as
exaggerated and kitsch in a painting and corny in
photographs; kind of like you just had to be there.
The breathing disorder was not asthma, after all,
but stress-related anxiety attacks. Broke up
with the boyfriend, and I'm breathing fine.

MEETING PLACE
1/10/98 — 12" x 8.25"

I was here a week before
to a dance. Then the place
was packed, and
surrounded by cars.
Tonight it's eerie and
quiet and deserted, so
unlike it was last week.
I had wanted to be
accompanied to this place,
but nobody was available.
So I'm here alone and it's
absolutely freezing.

LEAVING RON'S
1/11/98 — 7˝ x 8.5˝

This was done in my studio, not in plain freezing freak'n air. You know, I try to be a purist about these things, but I'm not a Viking or a polar bear. I'm not even much of a northerner anymore. There is a big difference between the memory images and those done *en plein air*. Outside images are fresher, take less time; studio-generated images are tighter, softer, more standardly lyrical.

Sky, Clouds, Moon
1/11/98 — 6″ x 6″

The title says it all. From memory.

MOONRISE
1/11/98 — 8.5″ x 12″

At the quarry with Ron. The sun sets as the moon rises and the chill of the evening air begins to settle into our bones. A different palette altogether from night painting. Two beavers come out and tussle with each other. They know Ron. They flatter him by ignoring his presence.

MOONSET
1/12/98 — 11″ x 9″

Ron phones at 5:45 a.m.
as planned. We climb in
moonlight, and descend
in daylight an hour later.
Bundled in many layers,
I am not uncomfortable
and manage to do this
painting entirely on
location. Back in the
studio, I saw off an inch
from the right vertical
side to make the image
stronger. It's rare that I
lop a chunk off a painting
like that, but painting on
wood panel makes it easy.

Natural & Unnatural Lighting
1/13/98 — 9″ x 7″

Looking through the trees at the lights — a simple
theme and it works. It amazes me how I struggle
with some pieces and they never look like much.
Then others simply happen. Like this one.

LIVING ROOM VIEW
2/9/98 — 9˝ x 10˝

This view is from my grandparents' apartment over-looking Central Park. It is also the reason they bought the place. Every evening Grandpa waits for the moon to appear. First red and large on the horizon, it gets smaller as it rises and competes with the city lights.

BACK HOME   One thing I love most about going away is
2/10/98 — 6″ x 4.5″   coming home and finding everything okay.
This painting celebrates that.

ROMA IN THE FULL MOON
2/11/98 — 11″ x 9″

This is from memory. You see how I'm getting easier with that, as I just blurted it out here in print. Not from life, although Roma is very much alive. He may be a reincarnated pet from my past. After all, why did he come here? Why did he stake the place out? Why did he refuse to leave after I placed him in a good home? An animal behaviorist told me this stray could not be given away successfully.

EVENING FLIGHT
2/11/98 — 13″ x 8.5″

Part memory, part
imagination. Weddings
and funerals have taken
precedence this month.
As I write this, the
painting is still not
completed. Grandmother
just died. What to put
flying across the sky?
I've not decided. Do I put
Grandma on a flying
carpet or in the clouds?
She would enjoy either.
I'm feeling too somber
to be whimsical.

ENROUTE TO QUARRY
2/11/98 — 10″ x 7″

Ron has been wonderful about going with me night-painting as long as our destination is a natural setting. It is freezing out so this was done in less than 15 minutes. I leave it on a log and retrieve it on our way back from the quarry.

9:30 P.M. AT THE QUARRY

2/11/98 — 9″ x 6″

A tree the beavers have spared works as a living moon-catcher. It is warmer down by the water, and so I am able to take more time with this painting. Ron wanders away to smoke a cigarette and not catch grief about it.

EDITH'S GARDEN
3/12/98 — 12″ x 8″

The cold is biting tonight.
Edith has over 100 garden
ornaments; the scene
deserves better detail but
my fingers are freezing, my
nose is running and I'll
simply have to come back
in spring.

SOLTERRA
3/13/98 — 9″ x 10″

Brian agrees to accompany me out to the development
where I bought land. He has put on a lot of clothes.
Indiana Jones hat, moon-walking boots, several layers
of pants and coats. Mercifully it's warmer than he
anticipated. Brian sits chatting away on a bunch of
topics, distracting me. So this painting takes much
longer than it should have, but I like the feel of it.

FOUR UGLY CARS
4/9/98 — 10″ x 11″

It's finally warmed up a little, leaves are popping out, with lots of help from rain, rain, rain.

4:30 am I awake and see the moon is out, so in my pjs, I paint this out the back door.

MARY'S POND
4/10/98 — 10˝ x 8.5˝

Too many obligations locally to go camping this full moon, so I'm taking the opportunity to socialize with friends. Mary Ellen hangs out with me as I do this rendition of her backyard. Her dogs, Monkey & Sam, occupy the space by day, running around the pond and in it. With them present it's impossible to paint here. They are into everything.

CATTAILS
4/17/98 — 8˝ x 5.5˝

These plants backlit from Mary Ellen's pond are memorable, so, alas, another image from memory.

Painting by Greta's place. Greta leaves to go to sleep. HOOT OWL DRIVE
Painting can be a lonely business. Cars go by at the   4/11/98 — 9˝ x 11˝
turn, back-lighting the trees. I'm glad that
not all roads are paved.

NEAR HOME
4/12/98 — 6.5″ x 5″

I'm wacko tired. I step
outside and do this with
the paint left on the
palette from the past
week. Lying in bed,
looking directly at the full
moon, I am drawn to paint
more, but my body just
won't let me.

New Leaves
4/21/98 — 12˝ x 8.5˝

Last fall I painted these trees in their raging red and
orange foliage. Six months later this image of them is
stronger, with their new green leaves budding. I've
balanced that wonderful fresh green with an orange
underpainting and some purple shadows.

THE CREEK IN NEW GREEN
$4/26/98$ — 12″ x 8.5″

Sitting here on the edge of a little wooden bridge as cars drive over, I can easily imagine bouncing direct, splash, into the water. A muskrat swims back and forth twice. She/he is indicated in the left corner.

AZALEAS
4/28/98 — 11″ x 8″

This is the main walkway to the Sarah P. Duke Gardens. The woman had such incredible insight, generosity and overall kindness to leave this garden legacy. She is someone I would like to have known.

HORSE WOMEN
5/9/98 — 10.5″ x 9.5″

Remembering I'm single, I kick myself out to circulate, art gear in tow, to a party. So what do I do? I spend the evening talking with two married women. They are both into horses and insist on this title.

GRANDPA'S FIRST SIGHTING

5/11/98 — 10.5″ x 8″

My brother phones. Grandpa is dying. His earliest memory was in the year 1900 — "I know I was still in a crib, I must have been very young. Very young. The curtain blew aside and there was this glowing ball. I was frightened. My mother came to my cries. She must have wondered, 'What the deuce is he crying about?', but I couldn't tell her. I couldn't speak yet." Ninety-plus years later he still laughs with wonder at his infancy dilemma.

DIRT ROAD
5/12/98 — 8″ x 6″

Back from working with a crew of non-painters, producing a mural that is, at best, floundering. I'm flopped down on the bed when the phone rings — Dave Owen. "Get outside. There's a great full moon." Ugh! I gather all supplies, trudge down the dirt road where I live and do this. I like it. Thanks, Dave. I love external motivation.

DAYLILIES
6/8/98 — 8″ x 6″

These flowers are
everywhere. 8 p.m. at
Dan's, they follow the
fading sun. It's hard to
render portraits of
flowering plants as
their time is fleeting,
but daylilies seem to go
for a good long stretch.

ROUTE 85
6/8/98 — 11″ x 9″

As I start this painting the moon is just up, but when I
get to adding it in, it's much higher in the sky.
Overpasses at night with huge trucks rumbling across;
lighted black monsters speeding through the night to
destinations unknown.

DAN'S BOUQUET
6/9/98 — 12″ x 9″

The almost-full moon is hidden tonight, so I'm immortalizing this bouquet Daniel presented me in his garden. The lightning bugs have hatched. I hold them in the same high esteem as praying mantises, butterflies, and luna moths.

FALLS OF THE NEUSE
6/10/98 — 6″ x 9″

When Kim bought this little place an hour's drive from where we live in Durham, I didn't get it. I thought, What a waste, but it turns out that this quiet waterfront retreat is like taking a vacation close to home. Very insightful and brilliant of her. I'm here for two nights.

SAME VIEW — DIFFERENT ANGLE
6/10/98 — 11″ x 8.5″

Kim sits happily watching the painting develop. She almost went to art school, but she is too practical and academically inclined. Meanwhile mosquitoes are biting only me so I'm painting very fast.

KIM BEFORE COFFEE
6/11/98 — 6″ x 8″

Waking up from my mattress on the lake-house floor, I see all these fabulous patterns. "You awake, Kim?" "Kind of." "Well, don't move for the next 20 minutes." Kim is delighted with becoming art first thing in the morning: "Amazing, that's me before coffee!!"

HANGOUT
6/11/98 — 9.5″ x 11″

It is completely overcast. I drive around the resort looking for other types of night light. I stop here and am asked to move on. I explain what I am doing and that I am not from the CIA or the FBI or there to bust anyone or anything. At midnight, packing to leave, I am invited to stay for dinner. "Blow" wanted to buy the painting on the spot.

BABY DUCKS
6/12/98 — 6˝ x 7˝

Kim's neighbors at the lake feed a particular mallard
duck. Recently she arrived with five ducklings.
The neighbor is very proud of this crew and feeds
them very special food. When he leaves for work
they will not be enticed by simple bread.
Alas, my models swim away.

WRONG WAY
6/25/98 — 9″ x 11″

Everything is growing well right now, weeds, everything. I love this peaceful, rural scene with a hint of danger.

VACANCY
7/6/98 — 7.5˝ x 6˝

Neighbors have moved
out. Hooray! Living next
to animal neglect is an
extreme challenge for me.
I hope wherever they go
they have more time for
their pets. It's been eerie
living down this dirt
road, all alone. I'm
packing now to go
to the beach.

SEAT BY THE SEA
7/8/98 — 11″ x 8″

I'm on the dock belonging
to a woman who
interprets handwriting.
We met briefly six years
ago, and here I sit
trespassing. She wouldn't
mind. When I return to
places like this, I'm
flooded with feelings,
thoughts and smells
from a forgotten past.

DUCK, ON HOLDEN BEACH
7/10/98 — 10″ X 11″

Mike was duck hunting Christmas Eve when he spied on a small island, in the middle of an icy river, a small, quivering puppy. Mike positioned himself level with his target, aiming his rifle just over the chunks of ice to end the young creature's misery. As he looked into its eyes he couldn't pull the trigger, but he cursed plenty as he tore off his clothes, plunging into the water and ultimately saving the little dog. In 17 years he has never regretted his decision. The dog he named "Duck" became his best friend.

DUKE CHAPEL
8/6/98 — 11″ x 8″

Back in Durham I go over to West Campus and set up in a dark corner of the churchyard. Several cats scatter. The patrolling security guard explains they are the abandoned pets of students. Good grief, I have a difficult time leaving my menagerie just for a weekend and with professional care! I don't get it; how do people do this?!

IRMGARD'S
8/7/98 — 6″ x 7.5″

Ron agrees to go searching with me for this magical place I stumbled upon years ago. We are far north of town, the sun has set and nothing matches my memory. We are passing a client's remote home, so I stop to say hi. Ron insists that at 9:30 p.m. and 15 years later, an impromptu reunion is a bad idea and rude. It's great to see Irmgard again and we talk like it was yesterday.

QUARRY
8/7/98 — 10.5˝ x 10˝

Back at Ron's I wander down to the quarry.
The plants are leafing out, scarcely resembling
the same scene I painted eight months earlier.

SAFE HAVEN
8/8/98 — 8″ x 11″

I'd planned on going camping in the mountains, but friends stopped me by recounting the current news of the mad bomber/killer on the loose up there. So here I am at Karin and Tripp's farm animal sanctuary, setting up in a big open field. I hear great hoofs approaching. I haven't gotten my night sight yet. Swoosh—hot air blows across my face. Surrounding me, I now see, is a herd of ex-veal calves, all over 1000 pounds each. The one closest, swooshing his great head, has long curling horns. I am terrified. He is now sniffing me, sniffing the paints, pawing the ground. Where is a tree? Do I play possum, run, grab his horns? Or just faint? I search my brain. Karin would not have a dangerous animal on her property; this is not Spain, I don't believe in bull fighting; I don't eat cows. I slowly pack up the supplies and creep back to the nearest tree, 50 feet away. The bull follows. I make it, but I'm thinking, The mountains couldn't be any more dangerous than this.

TABOULI
8/8/98 — 8″ x 6″

Karin joins me. The mosquitoes now have their complete battalion out. Tabouli stands directly in the moon's glow, posing magnificently. "Karin, is that great horned one dangerous?" "Deadly, he arrived with them, too old to dehorn, but because of his weapons he runs the place." Great. We go back to her straw-bale home to slather ourselves with bug repellent. My hands are swollen with bites. A mosquito remains stuck to the painting as testament to a major attack.

SHEEP & A GOAT
8/9/98 — 9.5″ x 10″

The donkey, Shady, places his head on my shoulder, the entire weight sinks in. He rubs his soft nuzzle against my face until I'm tumbled over laughing. Karin herds the donkey and dog distractions away. The cows are elsewhere. It is hard to make out the individual forms. I doubt this image will make sense. Amazingly, it does.

Audubon Lookout
8/10/98 — 11.5″ x 8″

Dave Owen, field
naturalist of the Eno
River, contacted me to
paint the moon from a
special point, preferably
at 3 a.m. Okay, so here we
are at 3 a.m.; fog is
thickening. This subtle,
calm image reminds me
of Dave. Feels in a way
like a portrait of him.

SUNFLOWERS
8/12/98 — 11″ x 8.5″

There's this garden at the summer camp I'm working at, with some of the most beautiful sunflowers in the world. This series was going to be completed in a year, but this is the 12th month since I started, and the country home of my recently deceased grandparents has yet to be featured. The year will have to be stretched one more month.

LAKE HOUSE
9/3/98 — 9″ x 10″

As mentioned, I'm extending the series to include my grandparents' country retreat. Here, my own folks awaited my arrival into the world. Every summer we visited and I hung out with the local hooligans.

MADELINE ROSE
9/4/98 — 7″ x 8″

My niece is night-painting with me. She and my sister are here almost every weekend, to escape the frenetic pace of the city. Maddie says she will be an artist when she grows up. Auntie Em suspects that with her skills in language, musical theatre and reasoning, the child has a shot at an exciting career with less crazy-making tendencies.

New Neighbors
9/4/98 — 10˝ x 7.5˝

For many years this
was a field of wheat.
The widow of the
naturalist who owned
it sold it to developers.
They built 70 houses.
We are no longer alone.

FROM THE
ROW BOAT
9/5/98 — 10˝ x 8˝

I cannot recommend painting from a row boat. This is the most challenging situation of the entire series (except for those that were life-threatening). Somehow, on a completely calm lake, this vessel is moving all over the place, so in one hand I've got a brush, in the other a freak'n oar, and I'm trying to row back into position. I view it as kind of a romantic experience. It isn't.

THE OLD WILLOW
9/6/98 — 9˝ x 12˝

Grandpa planted two willows in 1955 — they were
magnificent twins. This one passed on, but continues to
fertilize many others.

THE BOATHOUSE
9/7/98 — 8″ x 9″

It's got to be about 1 a.m. The moon's light has fully taken over and the colors I've mixed appear more naturalistic. I try not to use artificial light, and at this moment that purist yen presents no difficulty.

The Fire Pit
9/8/98 — 9″ x 6.5″

The moon is setting; it
has been raining all night.
As I burn the trash, I'm
brought back to when I
was a child and the night
closed in around us, and
we hung close to the fire.
We roasted marshmallows,
told ghost stories, acted
out skits and sang songs.
Why did that ever have
to stop?

PILOT MOUNTAIN MIST

10/5/98 — 10″ x 8″

Made up my mind to head out of town for the full moon, even though it was raining as far as I could drive. After a day and a night of hiking, painting, and camping in the rain with a dog that is acting miserable, I'm taking her sensible advice. We are heading home, but this place is really beautiful, even enveloped in a cloud.

EDITH'S CHAIR
10/7/98 — 12″ x 9.5″

Back at Edith's garden.
Edith has gone to bed.
It is past midnight and
my eyes are fully adjusted
to the night light. The
moonbeams are blasting.
The only light is the
moon's and nothing
impedes its brilliance.

ROMA AND DAISY
11/5/98 — 10.8″ x 8″

I'm relaxing about getting moon paintings done
as I kind of guess the series is over. This painting
celebrates the affection Roma continuously shows
Daisy. They take walks together several times a day.